T10669

KING'S CHRISTIAN SCHOOL
Christian Education Society of Enderby
350 B 30th St. N.E.
Salmon Arm, B.C. V1E 1J2
Tel: (250) 804-0340

TAKING OFF
PLANES THEN AND NOW

HERE WE GO!

TAKING OFF
PLANES THEN AND NOW

Steve Otfinoski

BENCHMARK BOOKS
MARSHALL CAVENDISH
NEW YORK

Benchmark Books
Marshall Cavendish Corporation
99 White Plains Road
Tarrytown, New York 10591-9001

Copyright © 1997 by Steve Otfinoski
All rights reserved. No part of this book may be reproduced in any form without written permission from the publisher.

Library of Congress Cataloging-in-Publication Data
Otfinoski, Steven.
Taking off : airplanes then and now / by Steven Otfinoski.
    p.   cm. — (Here we go!)
Includes bibliographical references and index.
ISBN 0-7614-0407-4 (lb)
1. Aeronautics—History—Juvenile literature. 2. Airplanes—Juvenile literature. I.Title II.Series: Here we go! (New York, N.Y.)
TL547.O85   1997   629.13'009—dc20   96-18677   CIP   AC

Photo research by Matthew Dudley

Cover photo: Courtesy of *U. S. Air Force Museum*

The photographs in this book are used by permission and through the courtesy of: *William B. Folsom:* 1, 15, 17, 26 (left and right), 27, 28. *The Image Bank:* Marc Solomon, 2; Magnus Rietz, 6; MarcRomanelli, 14; Gordon Kallio, 16; Williamson Edwards, 18–19; Gary Gladstone, 22 (top); Michael Melford, 24 (left), 25; Hank de Lespinasse, 29; Don Sparks, 30; Frank Whitney, 32; Reinhard Eisele, back cover. *Corbis–Bettmann:* 7, 8, 9, 10. *UPI/Corbis– Bettmann:* 11, 12, 13. *U. S. Air Force Museum:* 20–21. *Photo Researchers, Inc.:* Anthony Merciera, 22–23; Arthur H. Bilsten, 24 (right).

Printed in the United States of America

6 5 4 3 2

For William

Have you ever looked out of the window of an airplane after it takes off? The houses below look the size of match boxes and the people are as small as ants. Flying has been one of humankind's oldest dreams.

Italian artist and inventor Leonardo da Vinci drew a picture of a flying machine two years before Columbus sailed to America. To most people, da Vinci's invention was fantasy. How, they thought, could a person fly like a bird?

The first human flight was made in lighter-than-air balloons.
About one hundred years later, people built
winged gliders powered by the wind.
They looked remarkably like Leonardo's drawing.
Otto Lilienthal of Germany made two thousand short flights
with his gliders—his were the best.
He finally ran out of luck on August 9, 1896, and died when
his glider crashed.

The Wright Brothers in America studied Lilienthal's gliders.
They built an air machine made of wood and wire with two pairs of cloth wings.
A gasoline engine powered two propellers.
On December 17, 1903, at Kitty Hawk, North Carolina, they tested their flying machine.
Orville Wright manned the controls, while Wilbur Wright watched.
The plane flew 120 feet.
The flight lasted only twelve seconds, but in those seconds the age of the airplane was born.

Airplanes had improved a lot by the time World War I began in 1914.
Small fighter planes dueled with each other over the skies of Germany and France.
Here, a British pilot attacks German planes in a "dogfight."

After the war, many war pilots flew planes to entertain people at fairs and other events.
They swooped down out of the sky and flew upside down.
Many fliers walked on the wings in midair while a partner manned the controls.
These "barnstormers" risked their lives performing daredevil stunts.

Other young pilots joined the United States airmail service. They flew long hours in open cockpits through the rain and cold with no navigational equipment.
This is where the expression "flying by the seat of your pants" comes from.
One former airmail pilot, Charles Lindbergh (below), flew alone across the Atlantic Ocean to Paris, France, in 1927. He was hailed as a hero.

Another courageous early flier was Amelia Earhart (right), the first woman to cross the Atlantic alone. Earhart disappeared in the South Pacific while trying to fly around the world. What happened to her remains a mystery to this day.

In World War II, thousands of high-flying fighter planes filled the skies. Many took off from the decks of giant aircraft carriers.
This carrier-based F4U Corsair fighter (right) streaked through the sky at more than four hundred miles per hour.

The Stearman (above) was the plane new pilots learned to fly in.
It was painted yellow so people could see it easily—and stay out of its way.
People called the training plane the "Yellow Peril."

Bomber crews looked on their planes as old friends.
They gave them names and painted pictures on their sides.
Memphis Belle was one of the most famous American war planes.
A movie was made about her and her brave crew.

Huge bombers, like this B-25, dropped bombs over enemy territory.
Its cousin, the B-29, was called the "Superfortress" because of its size and flying power.
It flew long missions across the vast Pacific Ocean without needing to refuel.
Two B-29s dropped the first A-Bombs on Hiroshima and Nagasaki, two Japanese cities. The devastation led to Japan's surrender and the end of the war.

After the war, faster planes with jet engines were developed.
When the planes reached a speed of seven hundred miles per hour, something strange happened.
They shook and sometimes blew apart.
They had hit the sound barrier.
Engineers designed a new kind of jet, called *supersonic* (above sound).
The wings were bent back and the nose was long and pointed to "slice" through the sound barrier.
These new supersonic planes were named "X" for experimental.
An X plane like this one flew more than fifty miles above the earth in 1962.
The Jet Age had made way for the Space Age!

New passenger jets whizzed through the crowded skies. This Lear jet (left) flies business people to important meetings. The Concorde (below) was one of the first supersonic passenger airliners. A passenger can fly out of Paris in the morning and be in New York by lunchtime. The funny-looking bent nose allows the pilot to get a clear view of the runway when landing.

Small, single-engine planes are popular with amateur fliers.
Pilots can fly their families to their vacation homes or take their friends out for a spin.
This seaplane (right) has special floaters called pontoons that allow it to land on water.
Small planes can fly to remote areas where big planes can't go.
They can bring supplies, get people where they need to go, and help out in an emergency.

Modern fighter planes like the Fighting Falcon
(above left) helped the United States
win the Persian Gulf War in the Middle East in 1991.
The Galaxy (above right) carried soldiers and supplies
to the desert battlefields.
Stealth fighters (far right) are almost impossible
to pick up on radar because they fly so fast—up to
two thousand miles per hour.

This United States Air Force Thunderbirds unit is the last word in high performance precision flying.
The planes move together in formation like one giant machine.
The stunts they can do would make an old barnstormer jealous.

Where will the flying machines of the future take us?
The sky's the limit as planes take off into their second century.

INDEX

airmail 12
barnstormer 11, 29
bomber 18
Earhart, Amelia 13
fighter 10, 14, 26
glider 8
jet 23
Lilienthal, Otto 8
Lindbergh, Charles 12
single engine 24
sound barrier 21
Wright Brothers 9

FIND OUT MORE

Bellville, Cheryl Walsh. *The Airplane Book.* Minneapolis, MN: Carolrhoda Books, 1991.

Bendick, Jeanne, illustrated by Sal Murdocca. *Eureka! It's an Airplane!* Brookfield, CT: The Millbrook Press, 1992.

Berliner, Don. *Unusual Airplanes.* Minneapolis, MN: Lerner Publications, 1986.

Cave, Ron and Joyce. *What About Airplanes?* New York: Gloucester Press, 1982.

STEVE OTFINOSKI has written more than sixty books for children. He also has a theater company called *History Alive!* that performs plays for schools about people and events from the past. Steve lives in Stratford, Connecticut, with his wife and two children.

KING'S CHRISTIAN SCHOOL
Shuswap Christian Education Society
350 B 30th St. N.E.
Salmon Arm, B.C. V1E 1J2
Tel: (250) 832-5200